GAO

United States Government Accountability Office

Report to the Subcommittee on Energy and Water Development, Committee on Appropriations, U.S. Senate

November 2010

NUCLEAR WEAPONS

National Nuclear Security Administration's Plans for Its Uranium Processing Facility Should Better Reflect Funding Estimates and Technology Readiness

GAO-11-103

November 2010

NUCLEAR WEAPONS

National Nuclear Security Administration's Plans for Its Uranium Processing Facility Should Better Reflect Funding Estimates and Technology Readiness

Highlights of GAO-11-103, a report to the Subcommittee on Energy and Water Development, Committee on Appropriations, U.S. Senate

Why GAO Did This Study

Built in the 1940s and 1950s, the Y-12 National Security Complex, located in Oak Ridge, Tennessee, is the National Nuclear Security Administration's (NNSA) primary site for enriched uranium activities. Because Y-12 facilities are outdated and deteriorating, NNSA is building a more modern facility—known as the Uranium Processing Facility (UPF). NNSA estimates that the UPF will cost up to $3.5 billion and save over $200 million annually in operations, security, and maintenance costs. NNSA also plans to include more advanced technologies in the UPF to make uranium processing and component production safer.

GAO was asked to (1) assess NNSA's estimated cost and schedule for constructing the UPF; (2) determine the extent to which UPF will use new, experimental technologies, and identify resultant risks, if any; and (3) determine the extent to which emerging changes in the nuclear weapons stockpile could affect the UPF project. To conduct this work, GAO reviewed NNSA technology development and planning documents and met with officials from NNSA and the Y-12 plant.

What GAO Recommends

GAO is making five recommendations for, among other things, improving the UPF's cost and funding plans, ensuring that new UPF technologies reach optimal levels of maturity prior to critical project decisions, and for improving DOE guidance. NNSA generally agreed with the recommendations.

View GAO-11-103 or key components.
For more information, contact Gene Aloise at (202) 512-3841 or aloisee@gao.gov.

What GAO Found

The UPF project costs have increased since NNSA's initial estimates in 2004 and construction may be delayed due to funding shortfalls. NNSA's current estimate prepared in 2007 indicates that the UPF will cost between $1.4 and $3.5 billion to construct—more than double NNSA's 2004 estimate of between $600 million and $1.1 billion. In addition, costs for project engineering and design, which are less than halfway completed, have increased by about 42 percent—from $297 to $421 million—due in part to changes in engineering and design pricing rates. With regard to the project's schedule, NNSA currently estimates that UPF construction will be completed as early as 2018 and as late as 2022. However, because of a funding shortfall of nearly $200 million in fiscal year 2011, NNSA officials expect that the UPF will not be completed before 2020, which could also result in additional costs.

NNSA is developing 10 new technologies for use in the UPF and is using a systematic approach—Technology Readiness Levels (TRL)—to gauge the extent to which technologies have been demonstrated to work as intended. Industry best practices and Department of Energy (DOE) guidance recommend achieving specific TRLs at critical project decision points—such as establishing a cost and schedule performance baseline or beginning construction—to give optimal assurance that technologies are sufficiently ready. However, NNSA does not expect all 10 new technologies to achieve the level of maturity called for by best practices before making critical decisions. For example, NNSA is developing a technology that combines multiple machining operations into a single, automated process—known as agile machining—but does not expect it to reach an optimal TRL until 18 months after one of UPF's critical decisions—approval of a formal cost and schedule performance baseline—is made. In addition, DOE's guidance for establishing optimal TRLs prior to beginning construction is not consistent with best practices or with our previous recommendations. As a result, 6 of 10 technologies NNSA is developing are not expected to reach optimum TRLs consistent with best practices by the time UPF construction begins. If critical technologies fail to work as intended, NNSA may need to revert to existing or alternate technologies, possibly resulting in changes to design plans and space requirements that could delay the project and increase costs.

Changes in the composition and size of the nuclear weapons stockpile could occur as a result of changes in the nation's nuclear strategy, but NNSA officials and a key study said that the impact of these changes on the project should be minor. For example, the New Strategic Arms Reduction Treaty signed in April 2010 by the leaders of the United States and Russia would, if ratified, reduce the number of deployed strategic warheads from about 2,200 to 1,550. According to NNSA officials, NNSA and DOD have cooperated closely and incorporated key nuclear weapons stockpile changes into UPF's design. Also, an independent study found that most of the UPF's planned space and equipment is dedicated to establishing basic uranium processing capabilities that are not likely to change, while only a minimal amount—about 10 percent—is for meeting current stockpile size requirements.

United States Government Accountability Office

Contents

Letter		1
	Background	3
	UPF Project Costs Have Increased Since Initial Estimate and Construction May Be Delayed	6
	NNSA Is Developing Several New Technologies for the UPF and Is Assessing Their Maturity but Cannot Be Certain That All Technologies Will Work as Intended	10
	Emerging Changes in the Composition and Size of the Nuclear Weapons Stockpile May Only Have a Minor Effect on the UPF	18
	Conclusions	20
	Recommendations for Executive Action	22
	Agency Comments and Our Evaluation	23
Appendix I	**Objectives, Scope, and Methodology**	25
Appendix II	**Definitions of Technology Readiness Levels**	28
Appendix III	**Comments from the National Nuclear Security Administration**	31
Appendix IV	**GAO Contact and Staff Acknowledgments**	33
Tables		
	Table 1: New Technologies NNSA Is Developing for the UPF	11
	Table 2: TRL Definitions	14
	Table 3: TRLs for 10 Technologies NNSA Is Developing for the UPF	17

Abbreviations

DOD	Department of Defense
DOE	Department of Energy
NASA	National Aeronautics and Space Administration
NNSA	National Nuclear Security Administration
START	Strategic Arms Reduction Treaty
TRL	Technology Readiness Level
UPF	Uranium Processing Facility

This is a work of the U.S. government and is not subject to copyright protection in the United States. The published product may be reproduced and distributed in its entirety without further permission from GAO. However, because this work may contain copyrighted images or other material, permission from the copyright holder may be necessary if you wish to reproduce this material separately.

United States Government Accountability Office
Washington, DC 20548

November 19, 2010

The Honorable Byron L. Dorgan
Chairman
The Honorable Robert F. Bennett
Ranking Member
Subcommittee on Energy and Water Development
Committee on Appropriations
United States Senate

The Y-12 National Security Complex (Y-12 plant), located in Oak Ridge, Tennessee, is the National Nuclear Security Administration's (NNSA) site for conducting enriched uranium activities, producing uranium-related components for nuclear warheads and bombs, and processing nuclear fuel for the Navy.[1] Built in the 1940s and 1950s, the uranium processing operations at the Y-12 plant are outdated and deteriorating. According to NNSA officials, upgrading the Y-12 plant and maintaining it over the long term would require costly investments. In addition, the nation's nuclear weapons stockpile is shrinking, which has reduced the need for high-capacity enriched uranium activities and nuclear weapons component production. Therefore, in 2004, NNSA decided to construct a more modern facility—known as the Uranium Processing Facility (UPF)—that will consolidate uranium activities at the Y-12 plant from about 800,000 to 350,000 square feet. NNSA estimates the new facility will cost as much as $3.5 billion but that it will save over $200 million annually in operations, security, and maintenance costs.

As NNSA consolidates its facilities, it plans to develop more advanced technologies to make uranium processing and component production safer, more effective, and more efficient. Uranium processing uses chemicals and other means to recover enriched uranium from disassembled components and other scrap or salvaged materials in NNSA's inventory for use as fuel for naval and research reactors and re-use in new or refurbished nuclear weapons components. Component

---

[1] NNSA, a separately organized agency within the Department of Energy, was created by the National Defense Authorization Act for Fiscal Year 2000, Pub. L. No. 106-65, sec. 3201 *et seq.* (1999), with responsibility for the nation's nuclear weapons, nonproliferation, and naval reactors programs. NNSA owns the buildings, equipment, and the components produced at the Y-12 plant, which is operated under contract to NNSA by Babcock & Wilcox Technical Services Y-12, LLC.

production includes enriched uranium metalworking and other processes to assemble new or refurbished nuclear weapons components. Uranium processing and component production also involve hazardous processes that could expose workers to radiation or other dangers. NNSA is developing new, more advanced uranium processing and component production technologies that it hopes will reduce these potential hazards, according to NNSA officials. However, the risks inherent in relying on new, experimental technologies could affect NNSA's ability to construct the UPF within established cost and schedule estimates.

NNSA's plans for the UPF have also been affected by changes in the composition and size of the U.S. nuclear weapons stockpile. Existing uranium processing and component production capabilities at the Y-12 plant were designed to meet the large-scale demand that existed during the Cold War. The end of the Cold War has led to large reductions in the number of nuclear weapons in the stockpile, reducing the demand for uranium processing and component production.

In this context, you asked us to review the UPF. Our objectives were to (1) assess NNSA's estimated cost and schedule for constructing the UPF; (2) determine the extent to which the UPF will use new, experimental technologies and any risks to the project's cost and schedule of replacing the existing, proven technologies; and (3) determine the extent to which emerging changes in the nuclear weapons stockpile could affect the UPF.

To assess NNSA's estimated cost and schedule for constructing the UPF, we reviewed NNSA and contractor documents describing the project's cost and schedule estimates and recent design-related cost and schedule performance as well as documents showing cost and schedule implications for the future. We also interviewed officials at NNSA's Y-12 Site Office and NNSA's contractor for the Y-12 plant—Babcock & Wilcox Technical Services Y-12, LLC. To determine the extent to which the UPF will use new, experimental technologies and how NNSA plans to mitigate any resultant risks, we reviewed agency and contractor documents, including NNSA technology readiness reports and an independent study examining technology-related project risks. In addition, to understand NNSA's technology development goals, progress, and obstacles, we interviewed key NNSA and Y-12 plant officials responsible for maturing critical UPF technologies. We also visited the existing uranium processing and component production facilities that will be replaced by the UPF and observed demonstrations of several of the new technologies being developed. To determine the extent to which emerging changes in the nuclear weapons stockpile could affect the UPF project, we reviewed agency and contractor documents describing the key factors NNSA considered in developing UPF design plans that meet stockpile requirements. To obtain an independent perspective on the UPF design plans and approach, we also talked with officials at the Los Alamos and Lawrence Livermore National Laboratories who design the enriched uranium components that are to be produced at the UPF. We also reviewed a key independent study and discussed report findings with the study's principal author on how NNSA's UPF design plans are integrated with nuclear weapon stockpile requirements and how emerging changes in the stockpile could affect the UPF project.

We performed our work between December 2009 and November 2010, in accordance with generally accepted government auditing standards. Those standards require that we plan and perform the audit to obtain sufficient, appropriate evidence to provide a reasonable basis for our findings and conclusions based on our audit objectives. We believe that the evidence obtained provides a reasonable basis for our findings and conclusions based on our audit objectives. Appendix I contains a detailed description of our scope and methodology.

Background

Construction of the Y-12 plant in Oak Ridge, Tennessee, began in 1943 as part of the World War II Manhattan Project. The plant's early mission included the processing of enriched uranium necessary for building nuclear weapons. Today, the Y-12 plant continues its mission as NNSA's primary facility in the nuclear weapons complex for producing enriched uranium components necessary for maintaining the U.S. nuclear weapons stockpile. In addition, the Y-12 plant is used for dismantling weapons components, storing and managing nuclear material suitable for nuclear weapons, and processing fuel for Naval and research reactors, among other things.

Currently, the Y-12 plant consists of a patchwork of facilities and equipment that are not always efficiently connected, requiring the transport of materials during processing and component production operations. According to NNSA documents, the workflow is inefficient and requires a significant number of security personnel to patrol a relatively large protected area. Moreover, because of age and facility deterioration, operations and maintenance costs are continually rising with frequent outages and interruption in work schedules. According to NNSA officials, the existing facilities also do not meet a number of significant regulatory and design standards that are either in place or projected to be in the near future. For example, these facilities do not meet current standards for protection against natural occurrences or fire. Furthermore, existing Y-12 plant facilities do not provide optimal worker safety and protection from exposure to radioactive materials, including uranium, and other hazardous materials. Although these facilities have had periodic upgrades, the equipment, buildings, and support utilities need to be modernized for the Y-12 plant to continue to meet its mission, according to NNSA officials.

NNSA plans to transfer much of the ongoing uranium processing work and uranium component production that is performed at existing facilities at the Y-12 plant to the UPF in order to continue to support the nation's nuclear weapons stockpile and provide uranium fuel to the U. S. Navy, among other things. The proposed UPF is to consist of a single, consolidated uranium processing and component production facility to encompass less than half the size of the existing Y-12 plant facilities. NNSA officials expect that a combination of modern processing equipment and consolidated operations at the UPF will significantly reduce both the size and cost of enriched uranium processing at the Y-12 plant. Specifically, the officials said that the more-efficient layout of the new facility and more-modern equipment will significantly reduce processing and production costs, including costs associated with facility and equipment maintenance and maintaining worker and environmental health and safety.

DOE Order 413.3A establishes a process for managing the department's major projects—including contractor-run projects that build large complexes that often house unique equipment and technologies. The order covers activities from identification of need through project completion.[2] Specifically, the order establishes five major milestones—or critical decision points—that span the life of a project. These critical decision points are:

- Critical Decision 0: Approve mission need.

- Critical Decision 1: Approve alternative selection and cost range.

- Critical Decision 2: Approve performance baseline.

- Critical Decision 3: Approve start of construction.

- Critical Decision 4: Approve start of operations or project completion.

---

[2]DOE Order 413.3A was approved in 2006 and changed in 2008. This order canceled DOE Order 413.3, which was issued in 2000.

Order 413.3A specifies the requirements that must be met, along with the documentation necessary, to move a project past each milestone. In addition, the order requires that DOE senior management review the supporting documentation and approve the project at each milestone. DOE also provides suggested approaches for meeting the requirements contained in Order 413.3A through additional guidance.

For years, DOE and NNSA have had difficulty managing their contractor-run projects. Despite repeated recommendations from us and others to improve project management, DOE and NNSA continue to struggle to keep their projects within their cost, scope, and schedule estimates. Because of DOE's history of inadequate management and oversight of its contractors, we have included contract and project management in NNSA and DOE's Office of Environmental Management on our list of government programs at high risk for fraud, waste, abuse, and mismanagement since 1990.[3]

In response to its continued presence on our high-risk list, DOE analyzed the root causes of its contract and project management problems in 2007 and identified several major findings.[4] Specifically, DOE found that the department:

- often does not complete front-end planning to an appropriate level before establishing project performance baselines;

- does not objectively identify, assess, communicate, and manage risks through all phases of project planning and execution;

- fails to request and obtain full project funding;

- does not ensure that its project management requirements are consistently followed; and

- often awards contracts for projects prior to the development of an adequate independent government cost estimate.

To address these issues and improve its project and contract management, DOE has prepared a corrective action plan with various corrective

---

[3]GAO, *High Risk Series: An Update,* GAO-09-271 (Washington, D.C.: January 2009).

[4]DOE, *Root Cause Analysis: Contract and Project Management* (Washington, D.C.: April 2008).

measures to track its progress.[5] The measures DOE is implementing include making greater use of third-party reviews prior to project approval, establishing objective and uniform methods of managing project risks, better aligning cost estimates with anticipated budgets, and establishing a federal independent government cost-estimating capability.

UPF Project Costs Have Increased Since Initial Estimate and Construction May Be Delayed

NNSA's current cost estimates for constructing the UPF are already more than double its initial estimate. Moreover, the $200 million estimated annual savings in operations, maintenance, and security costs may not begin to be realized until the transition between existing uranium processing facilities at the Y-12 plant and the new UPF is complete. Although NNSA's current estimate prepared in 2007 indicates that the UPF construction will be completed between 2018 and 2022, NNSA officials expect the UPF will not be completed before 2020 due to funding shortfalls.

UPF Project Design Costs Have Already Increased

NNSA's current estimate, which was prepared in 2007 at critical decision 1, indicates that the UPF will cost between $1.4 and $3.5 billion to construct. This is more than double NNSA's initial 2004 estimate that was prepared at critical decision 0 of between $600 million and $1.1 billion. Cost estimates for project engineering and design, which are less than halfway completed, have already increased by about 42 percent—from $297 to $421 million. According to UPF project officials, these increases are the result of, among other things, changes in engineering and design pricing rates.

In January 2010, we reported that NNSA's current cost estimate for the UPF that was prepared in 2007 at critical decision 1 did not meet all cost estimating best practices because it did not exemplify the characteristics of a high-quality cost estimate.[6] As identified by the professional cost-estimating community in our *Cost Estimating and Assessment Guide*, a high-quality cost estimate is credible, well documented, accurate, and comprehensive.[7] However, our January 2010 report found that the UPF's

---

[5]DOE, *Root Cause Analysis Contract and Project Management Corrective Action Plan* (Washington, D.C.: 2008).

[6]GAO, *Actions Needed to Develop High–Quality Cost Estimates for Construction and Environmental Cleanup Projects*, GAO-10-199 (Washington, D.C.: Jan. 14, 2010).

[7]GAO, *Cost Estimating and Assessment Guide, Best Practices for Developing and Managing Capital Program Costs*, GAO-09-3SP (Washington, D.C.: March 2009).

current cost estimate prepared in 2007 only partially or somewhat met these four characteristics. For example, we found the UPF cost estimate only somewhat credible because an independent cost estimate had not been conducted. Instead, the project received an independent cost review as part of an independent technical review. An independent cost review is less rigorous than an independent cost estimate because it only addresses the cost estimate's high-value, high-risk, and high-interest aspects without evaluating the remainder of the estimate. Moreover, we found the UPF cost estimate was only somewhat accurate because it was not based on a reliable assessment of costs most likely to be incurred. The UPF cost estimate used an estimating methodology that was not appropriate for a project whose design was not stable and that was still anticipated to change. NNSA's technical independent review of the UPF stated that the project's cost-estimate range was unsupported in part because it was prepared with significant detail—for example, the estimate provided a count of pipings and fittings for the facility—despite the fact that there had been no design of technical systems or of the building on which to base these details. Our January 2010 report recommended, among other things, that DOE follow best practices and conduct an independent cost estimate for all major projects.

In response to our recommendation and recent congressional committee direction, DOE's Office of Cost Analysis is conducting an independent cost estimate on the UPF project before critical decision 2—approval of a formal cost and schedule performance baseline. This independent cost estimate is expected to be completed by the end of 2010. While this independent cost estimate may be used by NNSA headquarters officials as part of its process for approving the project's performance baseline, it is uncertain the extent to which Y-12 officials will accept the independent cost estimate results as reliable. Specifically, NNSA Y-12 project officials told us that the independent cost estimate will be based, in large part, on a subjective assessment of the independent cost estimating team's past experiences on similar construction projects. This is in contrast to the cost estimate prepared by the UPF project that is based on a detailed breakdown of the estimated prices of labor and materials specific to the UPF. Project officials noted that DOE's Office of Cost Analysis currently has no formal process for reconciling the two estimates given their different approaches. However, officials from DOE's Office of Cost Analysis told us that the independent cost estimate will be compared to the scheduled work and construction requirements specific to the UPF to understand what assumptions and cost elements are causing differences, if any, between the two estimates. According to these officials, this comparison will enable them and NNSA Y-12 project officials to understand cost risks for the project and determine how to address these issues. In addition, DOE is in the process of developing draft policy that is expected to help establish requirements and responsibilities for developing cost estimates for programs and performing independent estimates for program and project cost estimates. However, the current version of the draft policy does not specifically address how differing cost estimates should be reconciled.

Estimated Savings from the New Facility May Not Begin to Be Realized until Several Years after the UPF Is Built

According to NNSA officials, efficiency gains resulting from consolidating facilities at the Y-12 plant are likely to result in a savings of about $200 million annually in operations, maintenance, security, and other costs. For example, NNSA estimates it will save $54 million annually from the large reduction in the UPF's security perimeter when compared to the security perimeter around existing uranium processing facilities at the Y-12 plant. NNSA estimates cost savings will also result from the smaller amount of hazardous and radioactive waste the UPF will generate as compared to the existing facilities.

However, these savings may not begin to be realized until the transition between existing uranium processing facilities at the Y-12 plant and the new UPF is complete because both may need to operate simultaneously for an indeterminate period until the old facilities are decontaminated and decommissioned. For example, the Y-12 plant may need to continue to maintain some security in and around the old uranium processing facilities for some time after the UPF is built and operating because significant quantities of enriched uranium could still be present in the old facilities' piping and processing equipment during decontamination and decommissioning. According to NNSA officials, security measures in the old facilities can be significantly reduced once enriched uranium inventories are transferred to the UPF. In addition, unknown quantities of hazardous and radioactive waste will continue to be generated during the cleanup of the old facilities—prior to demolishing them—that will need to be treated and disposed, and potentially secured.

Although NNSA Currently Estimates UPF Construction Will Be Completed between 2018 and 2022, Funding Shortfalls Could Result in Delays

NNSA's current estimate prepared in 2007 at critical decision 1 indicates that the UPF construction will be completed as early as 2018 and as late as 2022. However, NNSA officials currently expect the UPF will not be completed before 2020 due to funding shortfalls. We have previously reported on DOE's use of unrealistic funding estimates while establishing cost and schedule baselines—a risk that also applies to NNSA major construction projects.[8] In addition, as discussed earlier, DOE's own root cause analysis of its contract and project management problems found that the department, among other things, fails to request and obtain full project funding. Consistent with our prior work and DOE's analysis, a 2007 technical independent review on the UPF project found a large disconnect between the funding available in NNSA's annual spending plan and the assumed annual funding levels in the UPF cost estimate.[9] Specifically, the review found that planned funding levels for fiscal years 2006 through 2008 did not meet the funding needs for the amount of work planned for those years. Despite this early warning of funding risks, NNSA officials approved the initial project cost range a few months after this technical review.

Moreover, with the submission of the President's budget for fiscal year 2010, NNSA officials anticipate a funding shortfall of nearly $200 million in fiscal year 2011 between what NNSA estimated the UPF project needed and what NNSA included in its budget request to Congress. NNSA officials said that this shortfall will likely delay project milestones and ultimately delay the UPF's estimated project completion from as early as 2018 to at least 2020 or later. This delay could, in turn, increase project costs. Potential funding shortfalls in subsequent years have also been identified as an ongoing high risk by project officials, which could result in additional unknown project delays and cost increases.

To address this concern about funding shortfalls, NNSA requested an internal review in February 2010 to ensure that UPF project funding

---

[8] GAO, *Department of Energy: Major Construction Projects Need a Consistent Approach for Assessing Technology Readiness to Help Avoid Cost Increases and Delays*, GAO-07-336 (Washington, D.C.: Mar. 27, 2007); *Nuclear Waste: Action Needed to Improve Accountability and Management of DOE's Major Cleanup Projects*, GAO-08-1081 (Washington, D.C.: Sept. 26, 2008); and *Department of Energy: Actions Needed to Develop High-Quality Cost Estimates for Construction and Environmental Cleanup Projects*, GAO-10-199 (Washington, D.C.: Jan. 14, 2010).

[9] NNSA, *Technical Independent Project Review of the Uranium Processing Facility Project at the National Security Complex Y-12*, IMA-PM-801768-A046 (Washington, D.C.: Apr. 24, 2007).

expectations from fiscal years 2012 through 2016 are reasonable. According to NNSA's briefing on the results of the review, NNSA's funding analyses appears to have addressed only whether the project would likely be able to spend the funds it requests in fiscal years 2012 and 2013. Importantly, the analysis appears to be incomplete because it (1) covers only 2 years and (2) does not address whether NNSA can realistically provide needed UPF funding given other NNSA priorities, such as other construction projects that will compete for funds in the same years. For example, according to NNSA's Future Years Nuclear Security Program accompanying the DOE's fiscal year 2011 congressional budget request, NNSA expects to request about $305 million in fiscal year 2012 to fund the Chemistry and Metallurgy Research Facility Replacement project at the Los Alamos National Laboratory, while requesting about one-third that amount—about $105 million—for the UPF.[10] Without assurance that NNSA mission priorities and its funding plans have been closely aligned with the UPF project's assumed annual funding levels, the UPF's cost and schedule estimates may not be credible.

NNSA Is Developing Several New Technologies for the UPF and Is Assessing Their Maturity but Cannot Be Certain That All Technologies Will Work as Intended

NNSA is developing 10 new technologies to install in the UPF and is using a systematic approach to gauge their maturity; however, NNSA may lack assurance that all technologies will work as intended before making key project decisions in accordance with best practices and our prior recommendations. If critical technologies do not work as intended, project officials may have to revert to existing or alternate technologies, which may result in higher costs and schedule delays.

---

[10] NNSA's Future Years Nuclear Security Program, included with DOE's annual budget request to Congress, contains NNSA's estimates of the funding it expects to request for the next 5 fiscal years.

NNSA Is Developing 10 New Technologies for the UPF and Is Using a Systematic Approach to Gauge Their Maturity

NNSA is developing 10 advanced uranium processing and nuclear weapons component production technologies for the UPF that, according to NNSA officials, will be more effective and efficient than existing technologies and that will reduce the hazards workers face at the Y-12 plant. (See table 1.) NNSA uses both chemical and metalworking processes and technologies to perform its work in the existing aging facilities at the Y-12 plant. For example, NNSA uses chemicals and other means to recover enriched uranium from disassembled components and other scrap or salvaged materials in NNSA's inventory. Once the uranium is recovered, it can be transformed into other forms, including powder-like enriched uranium oxide or uranium metal suitable for storage. In addition, NNSA uses enriched uranium metalworking processes to, among other things, prepare new or refurbished nuclear weapons components. For example, metalworking processes can include heating the uranium into liquid form so it can be poured into casts to create a variety of needed components. Metalworking processes also include machining operations where the uranium metal is cut on special tools at high speeds to create needed enriched uranium shapes. However, existing technologies at the Y-12 plant have become outdated, resulting in lesser levels of efficiency than would be possible with newer technologies. Existing technologies also expose workers to greater hazards because, for example, current machining operations are largely exposed and not automated, placing operators in greater contact with hazardous and radioactive materials.

Table 1: New Technologies NNSA Is Developing for the UPF

Technology	Description
Microwave casting	A process that uses microwave energy to heat and cast uranium metal into various shapes.
Infrared heating	A process to preheat and soften uranium metal prior to other processing activities.
Alternate processing of pins	A process to form uranium metal into custom shapes.
Bulk metal oxidation	A process that converts bulk uranium metal to oxide.
UNH calcination	A process that converts impure solutions into a stable, storable condition.
Saltless direct oxide reduction	A process that converts uranium dioxide into metal.
Recovery extraction centrifugal contactors	A process that uses solvent to extract uranium for purposes of purification.
Agile machining	A system that combines multiple machining operations—for fabricating metal into various shapes—into a single process.

Technology	Description
Chip management	An automated process that reduces operator interactions with machining process and improves worker safety by minimizing exposure to radioactive metal chips. It is one of the multiple operations to be performed through agile machining.
Special casting	A custom process for casting uranium metal into various shapes.

Source: NNSA.

Among the new technologies NNSA is developing are new chemical processing technologies for the UPF to address problems associated with current chemical processing technologies. For example:

- *Bulk metal oxidation.* This new technology for converting bulk uranium metal into a powder-like oxide will eliminate some intermediate processing steps in use at the Y-12 plant. The technology is expected to reduce the size of facilities needed for chemical processing and lessen workers' exposure to radiation and other hazards, among other things.

- *Saltless direct oxide reduction.* This new technology is expected to convert uranium dioxide into uranium metal, which would eliminate the use of some materials and processes that NNSA considers potentially hazardous to workers.

NNSA also plans to develop new metalworking technologies to produce uranium-related components at the UPF, including:

- *Microwave casting.* This technology uses microwave energy to heat uranium metal so that it can be poured into molds to produce various forms. It will replace an existing heating and casting process and is expected to be more effective, cost less to operate, and reduce the operator's exposure to uranium, according to NNSA officials.

- *Agile machining.* This technology consists of a system that combines multiple machining operations into a single, automated process. This new process is expected to improve worker safety by minimizing exposure to radioactive metal particles because all of the work will be performed within a sealed enclosure called a glovebox.

- *Chip management.* Among one of four subsystems of agile machining, NNSA is developing this technology as another means to achieve improved worker safety. For example, the new technology will replace manual operator tasks with a process that automatically collects uranium shavings, or chips. NNSA hopes this technology will help to minimize operator exposure to uranium.

Over the past several years, we have stressed the importance of assessing technology readiness to complete projects successfully, while avoiding cost increases and schedule delays.[11] Specifically, in 1999 and 2001, we reported that organizations using best practices recognize that delaying the resolution of technology problems until construction can result in at least a 10-fold cost increase. We also reported that an assessment of technology readiness is even more crucial at critical decision points in the project, such as approving a formal cost and schedule performance baseline, so that resources can be committed toward technology procurement and facility construction. Proceeding through these critical decision points without a credible and complete technology readiness assessment can lead to problems later in the project because the early warning of potential upcoming technology difficulties it provides would not be available to project managers.

To ensure that the UPF's new technologies are sufficiently mature in time to be used successfully, NNSA is using a systematic approach—Technology Readiness Levels (TRL)—for measuring the technologies' technical maturity. TRLs were pioneered by the National Aeronautics and Space Administration (NASA) and have been used by the Department of Defense (DOD) and other agencies in their research and development efforts for several years. DOE and NNSA adopted the use of TRLs agencywide in response to our March 2007 report that recommended that DOE develop a consistent approach to assessing technology readiness.[12] As shown in table 2, TRLs are assigned to each critical technology on a scale from a TRL 1, which is the least mature, through TRL 9—the highest maturity level where the technology as a total system is fully developed,

---

[11]GAO, *Best Practices: Better Management of Technology Development Can Improve Weapon System Outcomes*, GAO/NSIAD-99-162 (Washington, D.C.: July 30, 1999); *Joint Strike Fighter Acquisition: Mature Critical Technologies Needed to Reduce Risks*, GAO-02-39 (Washington, D.C.: Oct. 19, 2001); *Department of Energy: Major Construction Projects Need a Consistent Approach for Assessing Technology Readiness to Help Avoid Cost Increases and Delays*, GAO-07-336 (Washington, D.C.: Mar. 27, 2007); and *Coal Power Plants: Opportunities Exist for DOE to Provide Better Information on the Maturity of Key Technologies to Reduce Carbon Dioxide Emissions*, GAO-10-675 (Washington, D.C.: June 16, 2010).

[12]GAO, *Department of Energy: Major Construction Projects Need a Consistent Approach for Assessing Technology Readiness to Help Avoid Cost Increases and Delays*, GAO-07-336 (Washington, D.C.: Mar. 27, 2007).

integrated, and functioning successfully in project operations. Appendix II provides additional detailed information on TRLs.

Table 2: TRL Definitions

TRL	Definition
TRL 1	Basic principles observed
TRL 2	Concept/applications formulated
TRL 3	Proof of concept
TRL 4	Validated in a lab environment
TRL 5	Validated in a relevant environment
TRL 6	Subsystem demonstrated in a relevant environment
TRL 7	Subsystem demonstrated in an operational environment
TRL 8	Total system tested and demonstrated
TRL 9	Total system used successfully in project operations

Source: GAO analysis of DOD, NASA, and DOE data.

According to best practices we identified in our 2007 report, TRLs are useful because they:

- provide project managers with a method for measuring and communicating technology maturity levels from a project's design to its construction;

- provide a common language for project stakeholders, revealing any gaps between a technology's current and needed readiness;

- assist in decision-making and ongoing project management;

- increase the transparency of risk acceptance to identify technologies that most need resources and time; and

- reduce the risk of investing in technologies that are too immature.

NNSA Will Not Have Optimal Assurance That All Technologies Will Work as Intended before Reaching Key Project Dates

NNSA has made progress using TRLs to gauge the maturity of critical new UPF technologies; however, based on discussions with NNSA and contractor officials and our analysis of NNSA documents, NNSA does not expect to have optimal assurance as defined by best practices that 6 of the 10 new technologies being developed for UPF will work as intended before key project decisions are made. According to best practices we identified in our 2007 report, achieving an optimal level of assurance—

reaching specific TRL levels to provide assurance that the technologies will work as intended—prior to making critical decisions can mitigate the risk that new or experimental technologies will not perform as intended, which can result in costly design changes and construction delays.[13]

DOE's guidance on the use of TRLs recommends that new technologies achieve a TRL 6—the level where a prototype is demonstrated in a relevant or simulated environment and partially integrated into the system—by the time of critical decision 2—approval of a formal cost and schedule baseline for the project.[14] This is consistent with practices of other federal agencies such as the Department of Defense (DOD).[15] Most of the technologies NNSA is developing are expected to reach TRL 6 or higher by the time NNSA approves a formal cost and schedule performance baseline for installing this equipment in the UPF in July 2012.[16] For example, the new microwave casting technology is already at TRL 7. According to NNSA officials, NNSA has recently installed microwave casting technology in existing facilities at the Y-12 plant to demonstrate that it will heat enriched uranium as designed in an actual operational environment. As a result, NNSA will have high assurance that this technology will work as intended prior to approving the UPF's formal cost and schedule performance baseline.

However, NNSA does not expect to achieve the required levels of readiness for another key technology. Specifically, based on discussions with NNSA and contractor officials and our analysis of NNSA documents, NNSA does not expect one critical technology it is developing—agile machining—to reach TRL 6 until 18 months after approval of the project's cost and schedule performance baseline. Nevertheless, NNSA plans to approve its performance baseline with less than optimal assurance that this technology will work as intended. NNSA officials told us they have developed plans to address risks resulting from this technology readiness gap. Specifically, NNSA developed a technology maturation plan in early

---

[13] GAO-07-336.

[14] DOE, *Technology Readiness Assessment Guide*, DOE G 413.3-4 (Washington, D.C.: Oct. 12, 2009).

[15] DOD, *Technology Readiness Assessment (TRA) Deskbook* (Washington, D.C.: May 2005).

[16] As discussed above, funding shortfalls may result in project delays. According to UPF project progress reports, NNSA plans to update the estimated date for approving a cost and schedule performance baseline after it adjusts UPF project plans to account for, among other things, estimated shortfalls in fiscal year 2011 funding.

2010 to track technology development and engineering activities needed to bring the agile machining technology to TRL 6.

DOE's guidance on the use of TRLs is inconsistent with best practices used by DOD and with our previous recommendations with regard to technology readiness at another critical decision—start of construction. Specifically, DOD recommends that technologies reach TRL 7—the level where a prototype is demonstrated in an operational environment—prior to beginning its production and deployment phase, or the equivalent of beginning construction on a DOE project. Similarly, in 2007, we recommended that DOE construction projects demonstrate TRL 7 or higher before construction. Reaching this level indicates that the technology prototype has been demonstrated in an operating environment, has been integrated with other key supporting subsystems, and is expected to have only minor design changes. Nevertheless, DOE's guidance does not require technologies to advance from TRL 6 to TRL 7 between the approval of a formal cost and schedule baseline and the beginning of construction. Six of the 10 technologies NNSA is developing are not expected to reach TRL 7 before UPF construction begins. In the case of agile machining technology, NNSA expects that the technology will have only achieved a TRL 6 by December 2014 by the time of its expected procurement—1 full year after construction of the UPF is expected to begin in December 2013.[17]

Table 3 provides details on the current TRL for the 10 technologies, the TRL expected by the approval of a formal cost and schedule baseline in July 2012, the TRL expected by the start of construction in December 2013, and whether the expected TRLs meet best practices.

---

[17]As discussed above, funding shortfalls may result in project delays. According to UPF project progress reports, NNSA plans to update the estimated date for beginning UPF construction after it adjusts UPF project plans to account for, among other things, estimated shortfalls in fiscal year 2011 funding.

Table 3: TRLs for 10 Technologies NNSA Is Developing for the UPF

UPF technology	TRL—as of October 2010	TRL expected prior to formal approval of performance baseline	Does TRL expected at baseline approval meet TRL 6 as recommended by best practices?	TRL expected prior to construction start	Does TRL expected at start of construction meet TRL 7 as recommended by best practices?
Microwave casting	7	7	Yes	9	Yes
Infrared heating	7	7	Yes	7	Yes
Alternate processing of pins	7	7	Yes	7	Yes
Bulk metal oxidation	7	7	Yes	7	Yes
UNH calcination	5	6	Yes	6	No
Saltless direct oxide reduction	6	6	Yes	6	No
Recovery extraction centrifugal contactors	5	6	Yes	6	No
Agile machining	5	5	No	6	No
Chip management	5	6	Yes	6	No
Special casting	3	6	Yes	6	No

Source: GAO analysis of NNSA data.

Because all of the technologies being developed for the UPF will not achieve optimal levels of readiness prior to project critical decisions, NNSA may lack assurance that all technologies will work as intended. This could force the project to revert to existing or alternate technologies, which could result in design changes, higher costs, and schedule delays. In addition, other problems have occurred. For example, NNSA recently downgraded special casting technology from TRL 4 to TRL 3 because, according to UPF officials, unexpected technical issues occurred that required additional research and testing to resolve. Although officials expect this technology to be at TRL 6 by the time a formal cost and schedule baseline is approved in July 2012, it is not expected to reach TRL 7 before construction begins in December 2013.

A June 2010 NNSA management review of the UPF also noted that continued demonstration and testing of UPF technologies is still necessary.[18] The review stated that, because current operations in the Y-12

---

[18]NNSA, *Reasonableness Reviews, Chemistry and Metallurgy Research Replacement (CMRR) Project & Uranium Processing Facility (UPF) Project* (Washington, D.C.: June 3, 2010).

plant are expected to continue for over a decade longer, there appears to be a significant opportunity to demonstrate and test new technologies in an integrated fashion in the existing facility prior to installing them in the new facility. The review also noted that, if some technologies do not work as intended, it is not clear whether the current UPF design can accommodate the only identified alternative—to revert back to existing technologies. Furthermore, it noted that even with significant additional UPF investment, modifying the UPF's design could further delay the project. In such an event, the review concluded that continued operation of existing facilities at the Y-12 plant is NNSA's only strategy for addressing such delays.

Emerging Changes in the Composition and Size of the Nuclear Weapons Stockpile May Only Have a Minor Effect on the UPF

According to NNSA officials and an independent study commissioned by NNSA, emerging changes in the composition and size of the nuclear weapons stockpile as a result of changes in the nation's nuclear strategy or a proposed arms treaty with Russia should have relatively minor effects on the UPF project. The UPF's design is based on ensuring the facility has (1) sufficient capability—the space and equipment necessary to process enriched uranium and to produce the specific components for each type of weapon in the stockpile; and (2) sufficient capacity—the space and equipment necessary to produce the required quantities of components for the stockpile. As such, the elimination of a particular weapon type from the stockpile could eliminate some capability requirements in the UPF's design. Similarly, a reduction in the total number of weapons in the stockpile could reduce some capacity requirements in the UPF's design.

Changes in the composition and size of the stockpile could occur as a result of changes in the nation's nuclear strategy. Specifically, the April 2010 Nuclear Posture Review—the third comprehensive assessment of U.S. nuclear policy and strategy conducted since the end of the Cold War and conducted by the Secretary of Defense in consultation with the Secretaries of State and Energy—provides a roadmap for implementing the President's agenda for reducing nuclear risks and describes how the United States will reduce the role and numbers of nuclear weapons in the nation's nuclear security strategy, among other things.[19] For example, the review recommended studying the feasibility of using W-78 warheads that are currently used on intercontinental ballistic missiles on submarine-launched ballistic missiles. If this occurs, existing warheads used on

---

[19]DOD, *Nuclear Posture Review Report* (Washington, D.C.: Apr. 6, 2010).

submarine-launched ballistic missiles could be eliminated from the stockpile. According to the review, implementing the steps outlined in the report to reduce the role and numbers of nuclear weapons will take years and, in some cases, decades to complete.

In addition, the New Strategic Arms Reduction Treaty (New START) signed in April 2010 by the leaders of the United States and Russia would, if ratified, reduce the number of deployed strategic warheads from about 2,200 to 1,550. This treaty would replace the now-expired 1991 START I treaty and supercede the 2002 Strategic Offensive Reductions Treaty—also known as the Moscow Treaty—which expires in 2012. Further decreases in the size of the stockpile beyond those resulting from the New START treaty may also be possible. For example, the Nuclear Posture Review recommended a follow-on analysis to set goals for further warhead reductions.

NNSA officials told us that changes in the composition and size of the nuclear weapons stockpile should have relatively minor effects on the UPF project. Specifically, NNSA officials told us that they cooperated closely with DOD during the development of the Nuclear Posture Review and that several changes resulting from the review have already been incorporated into the UPF design. In particular, NNSA recently revised its primary project requirements document to accommodate expected changes in the composition and size of the nuclear weapons stockpile resulting from the Nuclear Posture Review and has already begun work to modify the UPF design to incorporate these changes. NNSA officials told us that changes made as a result of the close collaboration with DOD have helped to mitigate negative impact on the UPF project.

In addition, while NNSA has not formally studied the potential impact on the UPF if specific nuclear weapon types were eliminated, NNSA officials told us that such changes would likely not eliminate the need for capabilities currently designed into the UPF. Specifically, they said that if a warhead type were eliminated from the stockpile, the UPF's capabilities to produce a particular component for that specific warhead could potentially be eliminated from the project design. According to NNSA officials, because many of the UPF's capabilities will be used for common uranium chemical processing and component production operations, they therefore, are not limited to producing components for only one type of warhead. As a result, eliminating one type of warhead from the nuclear stockpile would not necessarily result in the elimination of a specific capability from the UPF's design because that capability could be needed for producing a wide range of other warhead types. For example, NNSA officials stated that replacing existing submarine-launched ballistic missile warheads with the W-78 intercontinental ballistic missile warhead would not significantly impact the UPF's design because this action would be unlikely to eliminate the need for equipment that is already planned to be installed in the UPF.

Moreover, an independent study commissioned by NNSA examining the UPF's space and major equipment needs concluded that changes in the size of the stockpile would result in relatively little change to the UPF's space and equipment design plans.[20] The study stated that establishing sufficient capability to meet minimum stockpile composition requirements—the ability to process enriched uranium and produce components for at least one of each weapon type in the stockpile—accounts for about 90 percent of the project's planned space and major equipment. Specifically, establishing minimum capabilities to, among other things, recover and process enriched uranium; produce, assemble, and dismantle nuclear weapons components; and produce fuel for naval nuclear reactors accounts for 91 percent of the facility's space and 89 percent of the UPF's major equipment. Only 9 percent of the UPF's space and 11 percent of the facility's major equipment are needed to ensure sufficient capacity to produce the necessary quantities of components to meet the requirements of the nuclear weapons stockpile. In other words, once the minimum capability is established, the overall impact on the project of modifying capacity to respond to changes in the size of the stockpile should be relatively minor. NNSA officials told us that adding or subtracting capacity can be addressed to a large degree by simply adding or subtracting work shifts on existing equipment.

Conclusions

When completed, the UPF will play an important role in ensuring the continued safety and reliability of the U.S. nuclear weapons stockpile. By replacing old, deteriorating, and high-maintenance facilities at the Y-12 plant, the UPF offers NNSA an opportunity to improve efficiency, save costs, and reduce hazards faced by workers at the plant. Because of its importance and given the size, scope, and expense of the project, it is critical that NNSA and Congress have accurate estimates of the project's costs and schedules. However, cost increases and potential schedule delays raise concerns about NNSA's ability to construct the facility within

---

[20]TechSource, Inc., *Independent Review of the Planned Space Design for the Uranium Processing Facility*, IMA-PM-801768-A082 (Germantown, Md.: Sept. 22, 2009).

its cost and schedule goals. In particular, NNSA's lack of a high-quality cost estimate for the project and its inability to consistently request and obtain sufficient project funding is consistent with the problems we discussed in our prior reports on DOE's difficulties in contract and project management, as well as the findings of DOE's own root cause analysis of this issue. NNSA is taking steps to provide independent assurance of the accuracy of its cost estimates for the UPF project. However, although DOE is developing draft cost estimating policy, NNSA lacks guidance for reconciling differences between the results of independent cost estimates and other project cost estimates. Moreover, NNSA's decision to approve an initial project cost range immediately after a 2007 technical review warned of a disconnect between the UPF project's funding requirements and NNSA's future years' spending plan, and then requesting $200 million less in fiscal year 2011 than the UPF project estimated it needed, raises concerns that NNSA is not placing sufficient high-level management focus on ensuring that UPF's cost and schedule estimates, and the associated funding plans these estimates are based upon, are consistent with NNSA's broader plans for funding the nation's nuclear weapons complex.

Managing a construction project of this type—particularly one that relies on several new or experimental technologies—is inherently challenging, and it is encouraging that NNSA is taking steps to manage the development of these technologies. For example, NNSA's early use of TRLs has already proven to be helpful in its efforts to mature these technologies. However, we are concerned because NNSA does not expect to achieve optimal assurance as defined by best practices that all 10 of these technologies will work as intended before key project decisions are made. Furthermore, because DOE's guidance for using TRLs is inconsistent with our prior recommendations as well as best practices followed by other federal agencies, DOE may be making critical decisions with less confidence that new technologies will work as intended than other agencies in similar circumstances. As a result, NNSA may be forced to modify or replace some technologies, which could result in costly and time-consuming redesign work. Moreover, Congress may not be aware that NNSA may be making critical decisions to proceed with construction projects without first ensuring that new technologies reach the level of maturity called for by best practices.

Recommendations for Executive Action

GAO is making five recommendations to improve NNSA's management of project funding and technology associated with the UPF project.

To improve DOE's guidance for estimating project costs and developing new technologies, we recommend that the Secretary of Energy take the following two actions:

- Include in the cost estimating policy currently being developed by DOE specific guidance for reconciling differences, if any, between the results of independent cost estimates and other project cost estimates.

- Evaluate where DOE's guidance for gauging the maturity of new technologies is inconsistent with best practices and, as appropriate, revise the guidance to ensure consistency or ensure the guidance contains justification why such differences are necessary or appropriate.

To improve NNSA's management of the UPF project, we recommend that the Secretary of Energy take the following three actions:

- Direct the Administrator of NNSA to ensure that UPF's cost and schedule estimates, and the associated funding plans these estimates are based upon, are consistent with NNSA's future years' budget and spending plan prior to approval of the UPF's performance baseline at critical decision 2.

- Direct the Administrator of NNSA to ensure new technologies being developed for the UPF project reach the level of maturity called for by best practices prior to critical decisions being made on the project.

- In the event technologies being developed for the UPF project do not reach levels of maturity called for by best practices, inform the appropriate committees and Members of Congress of any NNSA decision to approve a cost and schedule performance baseline or to begin construction of UPF without first having ensured that project technologies are sufficiently mature.

Agency Comments and Our Evaluation

We provided a draft of this report to NNSA for its review and comment. In its written comments, NNSA generally agreed with the report and our recommendations. NNSA stated that the UPF project is vitally important to the continued viability of NNSA's nuclear missions and is a top priority in its strategic planning efforts to transform outdated nuclear weapons infrastructure into a smaller, more modern nuclear security enterprise.

NNSA stated in its comments that its contractor has prepared an updated cost estimate that will be reflected in the President's fiscal year 2012 budget request and that independent cost estimates are being prepared in support of upcoming critical decisions for the UPF project. In addition, NNSA stated that it will work with DOE's Office of Engineering and Construction Management to ensure guidance on the reconciliation of cost estimates is incorporated in a new DOE cost estimating guide. Consistent with our recommendation, NNSA recognized in its comments the importance of having specific guidance on reconciling differences between the results of independent cost estimates and other project cost estimates.

Regarding its development of new technologies for the UPF, NNSA stated in its comments that our report does not discuss the risk management process used for the UPF project to manage technology risks and the many other risks for a project of this complexity and duration. NNSA is incorrect on this point. Our draft report discussed a number of steps NNSA is taking to mitigate technology risks. For example, our draft report noted that NNSA developed a technology maturation plan in early 2010 to track technology development and engineering activities needed to bring the agile machining technology to TRL 6.

NNSA also noted that TRL 6, as used by the UPF project in accordance with DOE guidance, has been judged to be an appropriate level of assurance that the technologies will work as intended when the final design of the project is complete and construction is ready to begin. Nevertheless, as our draft report noted, DOE's guidance on the use of TRLs is inconsistent with best practices used by DOD and with our previous recommendations with regard to technology readiness at the start of facility construction. Specifically, DOD recommends that technologies reach TRL 7—the level where a prototype is demonstrated in an operational environment—prior to beginning its production and deployment phase, or the equivalent of beginning construction on a DOE project. Similarly, we have previously recommended that DOE construction projects demonstrate TRL 7 or higher before construction. Reaching this level indicates that the technology prototype has been demonstrated in an operating environment, has been integrated with other key supporting subsystems, and is expected to have only minor design changes. However, DOE's guidance does not require technologies to advance from TRL 6 to TRL 7 between the approval of a formal cost and schedule baseline and the beginning of construction. Our recommendation that DOE evaluate its guidance to ensure conformance with best practices is intended to address these inconsistencies.

NNSA also provided technical comments that we incorporated in the report as appropriate. NNSA's written comments are presented in appendix III.

We are sending copies of this report to the appropriate congressional committees; Secretary of Energy; Administrator of NNSA; Director, Office of Management and Budget; and other interested parties. In addition, the report will be available at no charge on the GAO Web site at http://www.gao.gov.

If you or your staff have any questions about this report, please contact me at (202) 512-3841 or aloisee@gao.gov. Contact points for our Offices of Congressional Relations and Public Affairs may be found on the last page of this report. GAO staff who made major contributions to this report are listed in appendix IV.

Gene Aloise
Director, Natural Resources and Environment

Appendix I: Objectives, Scope, and Methodology

Our objectives were to (1) assess the National Nuclear Security Administration's (NNSA) estimated cost and schedule for constructing the Uranium Processing Facility (UPF) at the Y-12 National Security Complex in Oak Ridge, Tennessee; (2) determine the extent to which the UPF will use new, experimental technologies and any risks to the project's cost and schedule of replacing the existing, proven technologies; and (3) determine the extent to which emerging changes in the stockpile could affect the UPF project.

To assess NNSA's estimated cost and schedule for constructing the UPF, we visited the Y-12 plant and toured existing facilities as well as the proposed location of UPF. We also reviewed NNSA and contractor documents describing the project's cost and schedule estimates, budget documents, recent design-related cost and schedule performance, and documents potentially showing cost and schedule implications for the future. We also interviewed officials at NNSA's Y-12 Site Office and NNSA's contractor for the Y-12 plant—Babcock & Wilcox Technical Services Y-12, LLC.

To determine whether cost increases have occurred to date, we compared initial estimates for key activities, such as project engineering and design, with current estimates. We also obtained and reviewed NNSA documents describing the events that contributed to the cost increases, a Department of Energy (DOE) order on project management,[1] and a draft DOE order on cost estimating. We also used our January 2010 report that evaluated the UPF's cost estimates for compliance with industry cost estimating best practices.[2] We also obtained information on the independent cost estimate DOE's Office of Cost Analysis is conducting on the UPF project. Because NNSA's design of the UPF is less than halfway completed and because it has not yet established a formal cost and schedule performance baseline, current cost estimates are still considered to be preliminary and subject to change. Given this limitation, however, our analysis is meant to provide context for the condition of the current, pre-baselined cost and schedule estimate and to describe actions underway and planned to ensure the credibility of the formal cost and schedule performance baseline currently being developed.

---

[1]Department of Energy, Program and Project Management for the Acquisition of Capital Assets, DOE Order 413.3A (Washington, D.C.: July 28, 2006, and updated Nov. 17, 2008).

[2]GAO, *Actions Needed to Develop High–Quality Cost Estimates for Construction and Environmental Cleanup Projects*, GAO-10-199 (Washington, D.C.: Jan. 14, 2010).

Appendix I: Objectives, Scope, and Methodology

To determine the extent to which UPF will use new, experimental technologies and any risks to the project's cost and schedule of replacing the existing, proven technologies, we determined which critical technologies NNSA plans to use in UPF that are new or experimental. We visited the Y-12 plant to observe research and development activities associated with the technologies and reviewed agency and contractor documents, including NNSA technology readiness reports and an independent study examining technology-related project risks. In addition, we interviewed key NNSA and Y-12 plant officials responsible for developing UPF technologies.

To determine the extent to which NNSA was using industry best practices to ensure that new technologies will work as intended, we used best practices previously identified in our prior work and that are used by other federal agencies.[3] Specifically, best practices call for using a systematic method—Technology Readiness Levels (TRL), developed by the National Aeronautics and Space Administration (NASA) and used by other federal agencies such as the Department of Defense (DOD)—to determine the extent to which new technologies are sufficiently mature at key project decisions. TRL's use a scale to rate relative technology maturity on a scale from 1—being the least mature—to 9—representing the most mature ranking, where the technology has been demonstrated to work as intended in an operational environment. For each critical UPF technology, we obtained information from NNSA and UPF project officials on the current TRLs associated with each technology and compared them to optimal TRLs identified by best practices and DOE guidance on the use of TRLs. For technologies that are not expected to reach optimal TRL levels as identified by best practices and/or DOE guidance, we obtained information on NNSA's risk mitigation plans and its time frames for continuing research and development of the technologies. We also discussed with NNSA and UPF project officials the challenges that have been experienced or that they expect to encounter in the future. Finally, we compared NNSA's technology risk assessments with independent studies evaluating the maturity of planned UPF technologies.

---

[3]GAO, *Best Practices: Better Management of Technology Development Can Improve Weapon System Outcomes*, GAO/NSIAD-99-162 (Washington, D.C.: July 30, 1999); *Joint Strike Fighter Acquisition: Mature Critical Technologies Needed to Reduce Risks*, GAO-02-39 (Washington, D.C.: Oct. 19, 2001); and *Department of Energy: Major Construction Projects Need a Consistent Approach for Assessing Technology Readiness to Help Avoid Cost Increases and Delays*, GAO-07-336 (Washington, D.C.: Mar. 27, 2007).

Appendix I: Objectives, Scope, and Methodology

To determine the extent to which emerging changes in the stockpile could affect the UPF project, we visited the Y-12 plant and reviewed agency and contractor documents describing the key factors NNSA considered in developing the UPF's design in order to meet nuclear weapons stockpile requirements. In addition, we toured enriched uranium processing and nuclear weapons component facilities. We obtained the April 2010 Nuclear Posture Review issued by DOD and reviewed the proposed New Strategic Arms Reduction Treaty (New START) that was signed by the United States and Russia in April 2010. We also interviewed key NNSA and contractor officials to understand how changes in the composition and size of the nuclear weapons stockpile might affect the UPF's design. To ensure the reliability of the information we obtained from the UPF project officials, we obtained an independent perspective on the UPF's design through discussions with officials at the Los Alamos National Laboratory and Lawrence Livermore National Laboratory. These two nuclear weapons laboratories design the enriched uranium components that are currently produced at Y-12 and will be produced at the UPF. We also reviewed an independent study commissioned by NNSA examining the UPF's space and major equipment needs.[4] We met with the study's principal author and discussed the study's findings to determine how UPF's design is integrated with nuclear weapons stockpile requirements and how emerging changes in the stockpile could affect the UPF project.

We conducted this performance audit from November 2009 through October 2010 in accordance with generally accepted government auditing standards. Those standards require that we plan and perform the audit to obtain sufficient, appropriate evidence to provide a reasonable basis for our findings and conclusions based on our audit objectives. We believe that the evidence obtained provides a reasonable basis for our findings and conclusions based on our audit objectives.

---

[4]TechSource, Inc., *Independent Review of the Planned Space Design for the Uranium Processing Facility*, IMA-PM-801768-A082 (Washington, D.C.: Sept. 22, 2009).

Appendix II: Definitions of Technology Readiness Levels

Technology readiness level (TRL)	Level involved	Basic objective of TRLs	Components	Integration	Tests and environment
1. Basic principles observed and reported.	Studies.	Research to prove feasibility.	None.	None.	Desktop, "back of envelope" environment.
2. Technology concept and/or application formulated.	Studies.	Research to prove feasibility.	None.	Paper studies indicate components ought to work together.	Academic environment. The emphasis here is still on understanding the science but beginning to think about possible applications of the scientific principles.
3. Analytical and experimental critical function and/or characteristic proof of concept.	Pieces of components.	Research to prove feasibility.	No system components, just basic laboratory research equipment to verify physical principles.	No attempt at integration; still trying to see whether individual parts of the technology work. Lab experiments with available components show they will work.	Uses of the observed properties are postulated and experimentation with potential elements of subsystem begins. Lab work to validate pieces of technology without trying to integrate. Emphasis is on validating the predictions made during earlier analytical studies so that we're certain that the technology has a firm scientific underpinning.
4. Component and/or breadboard validation in lab environment.	Low fidelity breadboard.	Demonstrate technical feasibility and functionality.	Ad hoc and available laboratory components are surrogates for system components that may require special handling, calibration, or alignment to get them to function. Not fully functional but representative of technically feasible approach.	Available components assembled into subsystem breadboard. Interfaces between components are realistic.	Tests in controlled laboratory environment. Lab work at less than full subsystem integration, although starting to see if components will work together.

Appendix II: Definitions of Technology Readiness Levels

Technology readiness level (TRL)	Level involved	Basic objective of TRLs	Components	Integration	Tests and environment
5. Component and/or breadboard validation in relevant environment.	High fidelity bread/brass-board (e.g., nonscale or form components).	Demonstrate technical feasibility and functionality.	Fidelity of components and interfaces are improved from TRL 4. Some special purpose components combined with available laboratory components. Functionally equivalent but not of same material or size. May include integration of several components with reasonably realistic support elements to demonstrate functionality.	Fidelity of subsystem mock up improves (e.g., from breadboard to brassboard). Integration issues become defined.	Laboratory environment modified to approximate operational environment. Increases in accuracy of the controlled environment in which it is tested.
6. System/subsystem model or prototype demonstration in relevant environment.	Subsystem closely configured for intended project application. Demonstrated in relevant environment. (Shows will work in desired configuration.)	Demonstrate applicability to intended project and subsystem integration. (Specific to intended application in project.)	Subsystem is high fidelity functional prototype with (very near same material and size of operational system). Probably includes the integration of many new components and realistic supporting elements/subsystems if needed to demonstrate full functionality. Partially integrated with existing systems.	Components are functionally compatible (and very near same material and size of operational system). Component integration into system is demonstrated.	Relevant environment inside or outside the laboratory, but not the eventual operating environment. The testing environment does not reach the level of an operational environment, although moving out of controlled laboratory environment into something more closely approximating the realities of technology's intended use.

Appendix II: Definitions of Technology Readiness Levels

Technology readiness level (TRL)	Level involved	Basic objective of TRLs	Components	Integration	Tests and environment
7. Subsystem prototype demonstration in an operational environment.	Subsystem configured for intended project application. Demonstrated in operational environment.	Demonstrate applicability to intended project and subsystem integration. (Specific to intended application in project.)	Prototype improves to preproduction quality. Components are representative of project components (material, size, and function) and integrated with other key supporting elements/subsystems to demonstrate full functionality. Accurate enough representation to expect only minor design changes.	Prototype not integrated into intended system but onto surrogate system.	Operational environment, but not the eventual environment. Operational testing of system in representational environment. Prototype will be exposed to the true operational environment on a surrogate platform, demonstrator, or test bed.
8. Total system completed, tested, and fully demonstrated.	Full integration of subsystems to show total system will meet requirements.	Applied/integrated into intended project application.	Components are right material, size and function compatible with operational system.	Subsystem performance meets intended application and is fully integrated into total system.	Demonstration, test, and evaluation completed. Demonstrates system meets procurement specifications. Demonstrated in eventual environment.
9. Total system used successfully in project operations.	System meeting intended operational requirements.	Applied/integrated into intended project application.	Components are successfully performing in the actual environment—proper size, material, and function.	Subsystem has been installed and successfully deployed in project systems.	Operational testing and evaluation completed. Demonstrates that system is capable of meeting all mission requirements.

Source: GAO analysis of DOD data.

Appendix III: Comments from the National Nuclear Security Administration

Department of Energy
National Nuclear Security Administration
Washington, DC 20585

November 17, 2010

Mr. Gene Aloise
Director
Natural Resources and Environment
Government Accountability Office
Washington, DC 20548

Dear Mr. Aloise:

The National Nuclear Security Administration (NNSA) appreciates the opportunity to review the Government Accountability Office's (GAO) draft report, GAO-11-103, *NUCLEAR WEAPONS: National Nuclear Security Administration's Plans For its Uranium Processing Facility Should Better Reflect Funding Estimates and Technology Readiness*. We understand that the Subcommittee on Energy and Water Development, Senate Committee on Appropriations requested GAO to examine (1) NNSA's estimated cost and schedule for constructing Uranium processing Facility (UPF); (2) NNSA's plans for UPF integrate with nuclear weapon stockpile requirements; and (3) to what extent will UPF use new, experimental technologies and how does NNSA plan to mitigate resultant risks, if any?

The UPF project is vitally important to the continued viability of NNSA's nuclear missions and directly supports the Nation's deterrent strategy. The recently published Nuclear Posture Review validated this major system acquisition is a top priority in our strategic planning efforts to transform outdated Cold War legacy infrastructure into a smaller, more modern and efficient 21st century nuclear security enterprise.

We generally agree with the report and the recommendations. It should be noted that "current estimate," as used by the GAO throughout the report, refers to the cost estimate that was prepared in 2007 when Critical Decision 1 was achieved. Since the development of that estimate, the contractor has prepared an updated cost range estimate based on 45% design maturity, which will be reflected in the FY 2012 President's budget request. Consistent with project management best practices, NNSA will not set cost and schedule baselines until 90% design maturity.

Independent cost estimates are being prepared in support of the upcoming External Independent Review of the first Critical Decision 2/3 package; the establishment of the cost and schedule baseline for the Critical Decision 2/3 site work and Long Lead Engineered Equipment; and the CFO's Office of Cost Analysis has reviewed the project and is preparing a report on the overall estimated cost range of the UPF project. The reconciliation of cost estimates is considered a project management best practice. As such, NNSA will work with the Office of Engineering and Construction Management to ensure guidance on the reconciliation of cost estimates is

Appendix III: Comments from the National Nuclear Security Administration

incorporated and highlighted in the new DOE G 413.3-21, *Cost Estimating Guide*. These cost estimates and continuing reviews of the project will be considered when developing the final estimates of project costs.

Also, the report states that reaching "an optimal level of assurance," which is defined as Technical Readiness Level (TRL) 7 by best practices identified in a 2007 GAO report, prior to making critical decisions can mitigate the risk that new or experimental technologies will not perform as intended. However, the report does not discuss the Risk Management Process used for the UPF project to manage technology risks and the many other risks for a project of this complexity and duration. This is particularly true where the maturity of the *Advanced Integrated Machining System (AIMS)* equipment is a risk with a specific mitigation strategy. This process is being managed in parallel with the Technology Development and Maturation Plan on the UPF project. Additionally, TRL 6, as used by the Project in accordance with DOE G 413.3-4, *Technology Readiness Assessment Guide*, is judged to be an appropriate level of assurance commensurate with final design phase of the project and the minimum level where insertion or deployment is possible depending on technology complexity and risk acceptance by the stakeholders.

NNSA is committed to closely monitoring and balancing priorities with available resources to ensure the continued safe operations of the UPF. Enclosed are technical comments to help clarify and improve the report in areas that may be confusing or misleading.

If you have any questions related to this response, please contact JoAnne Parker, Director, Office of Internal Controls, at 202-586-1913.

Sincerely,

Gerald L. Talbot, Jr.
Associate Administrator
for Management and Administration

Enclosure

cc: Deputy Administrator for Defense Nuclear Programs
    Manager, YSO Site Office

Appendix IV: GAO Contact and Staff Acknowledgments

GAO Contact	Gene Aloise (202) 512-3841 or aloisee@gao.gov
Staff Acknowledgments	In addition to the contact named above, Ryan T. Coles, Assistant Director; John Bauckman; Virginia Chanley; Don Cowan; James D. Espinoza; Jonathan Kucskar; Alison O'Neill; Christopher Pacheco; and Tim Persons made key contributions to this report.

GAO's Mission	The Government Accountability Office, the audit, evaluation, and investigative arm of Congress, exists to support Congress in meeting its constitutional responsibilities and to help improve the performance and accountability of the federal government for the American people. GAO examines the use of public funds; evaluates federal programs and policies; and provides analyses, recommendations, and other assistance to help Congress make informed oversight, policy, and funding decisions. GAO's commitment to good government is reflected in its core values of accountability, integrity, and reliability.
Obtaining Copies of GAO Reports and Testimony	The fastest and easiest way to obtain copies of GAO documents at no cost is through GAO's Web site (www.gao.gov). Each weekday afternoon, GAO posts on its Web site newly released reports, testimony, and correspondence. To have GAO e-mail you a list of newly posted products, go to www.gao.gov and select "E-mail Updates."
Order by Phone	The price of each GAO publication reflects GAO's actual cost of production and distribution and depends on the number of pages in the publication and whether the publication is printed in color or black and white. Pricing and ordering information is posted on GAO's Web site, http://www.gao.gov/ordering.htm. Place orders by calling (202) 512-6000, toll free (866) 801-7077, or TDD (202) 512-2537. Orders may be paid for using American Express, Discover Card, MasterCard, Visa, check, or money order. Call for additional information.
To Report Fraud, Waste, and Abuse in Federal Programs	Contact: Web site: www.gao.gov/fraudnet/fraudnet.htm E-mail: fraudnet@gao.gov Automated answering system: (800) 424-5454 or (202) 512-7470
Congressional Relations	Ralph Dawn, Managing Director, dawnr@gao.gov, (202) 512-4400 U.S. Government Accountability Office, 441 G Street NW, Room 7125 Washington, DC 20548
Public Affairs	Chuck Young, Managing Director, youngc1@gao.gov, (202) 512-4800 U.S. Government Accountability Office, 441 G Street NW, Room 7149 Washington, DC 20548

www.ingramcontent.com/pod-product-compliance
Lightning Source LLC
Chambersburg PA
CBHW081245180526
45171CB00005B/543